TIMBER COUNTRY

BY LYNN M. STONE

THE ROURKE CORPORATION, INC.
Vero Beach, FL 32964

Edited by Sandra A. Robinson

PHOTO CREDITS

All photos © Lynn M. Stone except page 39 © Tom and Pat Leeson, page
31 © Weyerhaeuser Timber Company, and page 37 courtesy of the United
States Department of Agriculture-Forest Service

DEDICATION

For good friends in Timber Country, the Trumbos

Library of Congress Cataloging-in-Publication Data
Stone, Lynn M.
    Timber country / by Lynn M. Stone.
        p. cm. — (Back roads)
    Includes index.
    Summary: An introduction to the Pacific Northwest, a forested area
of great natural beauty, emphasizing the timber industry there.
    ISBN 0-86593-305-7
    1. Forests and forestry—Northwest, Pacific—Juvenile literature.
2. Timber—Northwest, Pacific—Juvenile literature. 3. Forest ecology—
Northwest, Pacific—Juvenile literature. 4. Logging—Northwest, Pacific—
Juvenile literature. [1. Northwest, Pacific—Description and travel.
2. Forests and forestry—Northwest, Pacific. 3. Timber.] I. Title. II. Series:
Stone, Lynn M. Back roads.
SD144.A13S76 1993
634.9'09795—dc20   93-4522
                                              CIP
                                              AC

TABLE OF CONTENTS

1. Timber Country 4

2. The Forests of
   Timber Country 13

3. From Forest to Home 23

4. Renewing the Forest 31

5. Who Owns the Timber
   in Timber Country? 34

6. Timber and the Environment ...38

7. Visiting Timber Country 42

   Glossary 45

   Index 47

CHAPTER 1
TIMBER COUNTRY

America's Pacific Northwest is timber country. Some of the greatest forests on earth grow here in western Oregon, western Washington and northern California.

Some of the greatest forests on earth cover much of the Pacific Northwest

Deep forests of Douglas-fir, Sitka spruce and western hemlock trees follow river valleys, hug mountain slopes, and step to cliff edges overlooking the Pacific Ocean. Forests of magnificent redwoods, the tallest trees on earth, are scattered along a narrow coastal band from west central California into extreme southern Oregon.

Nature was a big spender in the Northwest. She created a landscape of rocky seashores, thundering rivers, waterfalls, glaciers, mountain peaks and lakes. Few places on earth rival the natural beauty of timber country. The land still has its rural, even wilderness, character. Back roads slip into towering forests, wind into the valleys of wild rivers, and edge onto headlands by the sea.

Flying above timber country, one can travel for miles and see little else below except dark forest. The green forests of the coastal mountain ranges stand like wrinkles near the Pacific coast. Further inland, the Cascades in Oregon and Washington, and the Sierras in California are taller, more rugged mountains than the coastal ranges. Their lower slopes are forested, but their snowy crowns rise above **timberline,** the level at which trees can grow.

Here and there in timber country are breaks in the forests – dairy farms, orchards, fingers of the sea, dunes, and the cities and towns – like Seattle, Tacoma, Portland, Waldport, Bend, Olympia, Salem and hundreds of others. But it is the forests, not the breaks in the forest, that have put their stamp on the Pacific Northwest. Ever since

Mount Rainier, snow-crowned jewel
in the Cascades, rises above wildflower
meadows and stands of spruce

Dairy farms are nestled among the forested hills of the Tillamook valley in Oregon, where Tillamook cheese is a local specialty

American settlers began to pioneer the region more than 150 years ago, the Pacific forests have been filling a wide range of human needs. Forests have supplied the raw materials for the wood products used in this country and abroad. The forests of timber country have also provided

the jobs that yield wood products such as lumber for
construction and wood pulp for paper.

The cutting of trees and the processing of wood are
important sources of **revenue,** or money, in the Northwest.
Oregon, for example, is the leading timber-producing

state in the United States, which is the world's leader in timber production. In recent years Oregon has produced nearly 70 million cubic yards of wood annually. Washington is the third leading state in wood production, behind Georgia. California is ranked sixth among the states in wood production. The timber industry is no less important in British Columbia, Washington's Canadian neighbor to the north. British Columbia forests turned out over 97 million cubic yards of wood in a recent year.

In addition to supplying wood and jobs, the forests fulfill other important needs for humans and wildlife. Forests freshen air by releasing oxygen and removing carbon dioxide. They filter water that settles into the ground, and they act as sponges to reduce the likelihood of surface runoff and flooding. The roots of forest trees anchor soil. That prevents dirt from fouling rivers and streams and destroying the **habitats,** or living places of salmon and trout. The forests' hundreds of species of plants are habitats for an amazing variety of animals – from microscopic creatures and slimy banana slugs to owls, squirrels, cougars, black bears and Roosevelt elk. The forests are also habitats for the tens of thousands of people who enjoy hiking, camping, fishing, hunting and nature study in timber country.

Oregon forests produce more wood than any other forests in the United States

Finger-length banana slugs seem out-of-place in the forests of giant redwoods, but they thrive in the moist air

CHAPTER 2

THE FORESTS OF TIMBER COUNTRY

The forests of timber country are always green, barring fire or disease. This is because most of the trees in these forests continuously shed and renew their leaves, rather than lose them all at once as broad-leaved, **deciduous** trees do. The dominant trees in these forests are **conifers.** Conifers, such as pines, firs and spruce, bear cones and have needlelike leaves. The most important conifers in the region are the Douglas-fir, Sitka spruce, western hemlock and western red cedar.

Much of timber country is Pacific coastal forest. This is a type of forest that consists largely of certain evergreens. It stretches from west central California northward through the coastal ranges of Oregon, Washington, British Columbia and southeast Alaska. Further inland, away from the immediate influence of the Pacific Ocean, mountain evergreen forests dominate timber country. Several species of pines are among the important trees in these somewhat higher and drier forests of the Northwest.

The forests of timber country are dominated by needle-leaved evergreens – fir, spruce, cedar, pine and hemlock

Coastal forest thrives in mild air, gentle mist and rain. The mild climate is the product of warm ocean currents. Prevailing west winds blow clouds laden with Pacific moisture onto the mainland. Most of it falls in the form of rain, which drenches the coastal forests. Some of the coastal mountains of timber country are soaked by up to

130 inches of precipitation each year. The rainfall that pelts coastal Oregon, which may be the best timber-producing area in the world, is about three times the precipitation of Chicago or New York. But concrete and steel skyscrapers grow tall without the help of moisture. The timber skyscrapers need the rain.

Moisture sweeping ashore from the Pacific drenches the forests of the coastal ranges

On Washington's Olympic peninsula, rainfall sometimes exceeds 130 inches per year and creates a special variation of the coastal forest called **temperate** rain forest. The temperate rain forest should not be confused with the hot rain forests of the world's tropics. These northern rain forests of western Washington and British Columbia have essentially the same northwestern tree species as the rest of the coastal forest – Douglas-firs and Sitka spruces, for example. The temperate rain forests, however, are draped and insulated with beards and mats of moss. They are forests of green light, where moss clings to bark, branches and every rotting log.

Even more fascinating than the rain forests are northern California's redwood groves, another variation in the coastal forest. Some of these giant conifers stand over 300 feet tall, the length of a football field. On summer mornings their crowns disappear in the Pacific fog that drifts inland and helps the redwoods preserve their moisture.

Mosses carpet the woodland floor in this temperate rain forest along the Hoh River in Olympic National Park

The forests vary in age and in their composition of trees. Most of the forests in timber country are **second-growth** or "intermediate" forests. They are "second" forests, having grown up after the death of a previous forest on the same land. The original forest may have burned at some time in the last 250 years, or it may have been logged. Some of the remaining forests in timber country, however, including several of the last few redwood groves, are "old growth" or "ancient" forests. Old growth forests have essentially never been disturbed by logging, nor have they burned in the last 250 years. Some of these old kingdoms of trees are far more than 250 years old.

The old growth forests are remarkable not only for their age, but for their beauty and diversity as well. Old growth forests are more than just retirement villages for huge, aged trees. The ancient forests are the most complete communities of the forest world. Mature old trees stand among dying trees, young trees and centuries of fallen logs, branches and forest litter. Where old trees have collapsed, leaving light and space, young, vigorous trees replace them. Some of the older trees, split by lightning or wracked by disease, make ideal hosts for a variety of insects. Birds feed on the insects and nest in the tree hollows. The ancient forests have a well-developed understory of smaller trees and shrubs beneath the crowns of the mature trees. They also have a rich carpet of ferns and wildflowers.

California redwoods, the world's tallest trees, bathe their crowns in morning fog, a necessary element in their survival

A Douglas squirrel shreds bark for its nest in a Washington forest

The old growth forest has a **niche,** a proper place in the community, for each woodland creature. The welfare of the forest depends on the relationships of those creatures and the plants around them. Each member of the community is dependent on one or more of the other members. Basically, each plant and animal relies on other plants and animals for its food and survival.

Trees and other green plants use sunlight in their food-making process, which also involves the air, water and **nutrients,** or nourishing elements, in the soil. The plants and their seeds and fruits become food for various animals. Other animals, the forest **carnivores,** or meat eaters, feed on the plant eaters. Food energy passes

Light shafts brighten the old growth redwood forest in Redwood National Park

through many food chains that make up a complex forest food web. A simple chain within the web might start with a cone from one of the forest conifers. A Douglas squirrel eats the cone and thus gains some of the food energy that was produced and stored by the conifer. Eventually, an owl kills and eats the Douglas squirrel. Some of the energy stored in the squirrel is transferred to the owl. When the owl dies, its body is rapidly reduced by decay, the action of insects and tiny organisms known as bacteria, protozoans and fungi. This process helps return the owl and its food energy to the soil. There it can be used by plants to manufacture food and new plants, restarting the energy cycle.

CHAPTER 3

FROM FOREST TO HOME

Bicyclists and auto drivers in timber country are warned by frequent highway signs that logging trucks use the roads, too. No one wants to be caught on the wrong side of the road when logging trucks are rolling. Loaded up with trimmed, freshly-cut trees, some of the trucks are bound for seaside docks in places like Newport and Coos

A bulldozer helps build a trail for loggers who will thin second growth in this national forest

Bay, Oregon. From these ports, logs will be shipped overseas for further processing. Other trucks are rumbling to sawmills in timber country.

Nearly everyone who travels in the Pacific Northwest sees logging trucks. Few see the trees actually being cut. But freshly-cut logs don't suddenly appear like Jacob Marley's ghost in *A Christmas Carol*. It's just that the chances of seeing a timber operation are very small because timber country is very large. The vast majority of cuts do not occur along a paved, public road. Rather, the cuts are in back country. Anyway, an active logging area is no place for visitors. Logging is demanding and dangerous – strictly a hard-hat occupation. A logging site teams huge machines, tractors, trucks, powerful saws and long steel cables. A careless visitor – or worker – is no match for any of them.

Logging begins after a cutting site has been chosen and a dirt road to it has been opened. If the forest acres selected for cutting are to be **clear-cut,** all of the trees will be taken down. If the site is to be selectively cut, the forest will be thinned, and only hand-picked trees will be removed.

Trees are sawed by a faller. The faller is a modern-day lumberjack. He uses a growling chainsaw as easily as an artist uses a brush. However, he remembers that cutting trees is more like bullfighting than painting. A faller usually

Protected from flying wood chips by goggles, a tree faller saws through a slender Douglas-fir

25

knows in which direction a tree will fall, and how it will "jump" from the stump. But just as a bull can be unpredictable, so can a Douglas-fir.

A fallen log is removed from the forest immediately. On some sites, a machine called a cable yarder retrieves the log. A cable yarder spools out steel cable like a fishing reel. After workers attach the log to a **choker** on the cable, the yarder retrieves the cable and log. A **grapple loader** lifts the log from a stack next to the yarder and sets it onto a waiting log truck.

Cable yarders are especially useful on rugged slopes. In level forests, a **skidder** may be used to remove logs. A skidder is a bulldozer clone, with big rubber tires and steel jaws known as a grapple. The grapple clamps the log, and the skidder hauls it from the forest. On logging sites where suitable haul roads cannot, or should not, be built, trees are occasionally – and very expensively – lifted out by helicopter.

Logs are measured, numbered and rated for quality by a **scaler** at the sawmill or dock. Eventually the logs are sawn into boards or in some other way processed for sale.

A tree faller watches as a Douglas-fir pitches down, providing more light and growth opportunity for remaining trees

A grapple loader, steel arm of the cable yarder, uses sharklike jaws to hoist a log that will soon be placed on the log truck's bed

Wood is used for more products than most people know. Everyone is familiar with wooden furniture, plywood, fence posts, dock pilings and construction lumber. Wood is also used in the manufacture of particle board, paper, cardboard and insulation. Chemically processed wood is an ingredient in certain plastics, lacquers, dyes, adhesives, rayon, concrete and food products.

Logging crews brave brush, hornets and bulky cables to attach logs for retrieval by a cable yarder

The scaler will measure, record and rate each log

CHAPTER 4

RENEWING THE FOREST

This tree farm in timber county is being replanted by hand with seedlings

People who manage forests for the production of timber treat the forest like a crop for harvest. Like any crop, trees can be replanted and reharvested if they are managed wisely. Timber producers use the term "sustained yield" – the principle of cutting no more timber than is grown during any time period. Good forest managers keep more trees in the "bank" than are being spent at any one time.

Clear-cut land in an Oregon national forest will be cleared of debris and quickly replanted

After a forest has been cut for timber, the land will reforest itself. But reforestation by nature is a lengthy process, and during that period the land may lie barren, erode, and foul rivers and streams. The solution is to give nature a human hand and accelerate the reforestation.

Reforesting basically means to replant the forest, either by seeds or by young trees called **seedlings.** In a typical reseeding, foresters first remove tree debris from a cutover site. Seeds, as many as 30,000 per acre, or seedlings are then planted. Machines generally do the planting, but occasionally planting by hand is the preferred method.

In hard-to-reach areas, helicopters seed cutover lands. Douglas-fir and western hemlock are usually planted, but the choice of tree species often depends upon who is doing the planting, and what kind of trees had been on the land before. Foresters on private timber lands often plant only one species of tree on a site. Redwood tree farms in California plant that species exclusively. Foresters on public lands, such as America's national forests, are increasingly planting trees to reflect the former composition of the forest.

Forest renewal is the subject of considerable study. The science of harvesting and planting trees is called **silviculture.** Silviculturists work with experimental forests and in laboratories to find improved ways to grow trees and manage forests. Tree farmers working with silviculturists have learned, for instance, that redwoods can be grown on their tree farms at rates 10 times faster than they grew in natural forests.

CHAPTER 5

WHO OWNS THE TIMBER IN TIMBER COUNTRY?

About half of Washington and Oregon is covered by forest. Approximately 40 percent of California is forested. Who owns all the forest land?

Millions of forest acres are owned by private forest products companies, such as the Georgia Pacific Corporation, Weyerhaeuser Company and International Paper Company. Millions of additional acres are owned and managed by private tree farms. Most forest land in timber country, however, is owned by various governments and their agencies. The state governments of timber country and the provincial government of British Columbia own forest land. Several United States government agencies own forest lands, too, but the manager of the largest stands of timber in the Northwest is the U.S. Forest Service, an agency of the Department of Agriculture. Nationwide, the Forest Service manages 191 million acres of forest lands. Many millions of those green acres are in the 38 national forests of Oregon, Washington and California.

Insect-eating California pitcher plants, also known as cobra plants, are protected in the Siuslaw National Forest's Darlingtonia Preserve, a forest bog in Oregon

National forest signs advertise, quite accurately, that visitors are in a "Land of Many Uses." The Forest Service provides campgrounds, picnic areas, interpretive centers, trails and boat ramps for public use. It operates lookout towers and firefighting equipment. Forest Service biologists study the life of the forests and help decide how certain plants and animals should be managed. National forests protect a variety of natural wonders, from ancient rock formations and lava tree casts, to insect-eating California pitcher plants. On a grander scale, many of the national forests in timber country protect huge blocks of wild lands known as wilderness areas. Only foot traffic is permitted in the designated wilderness.

The national forests also work with private interests. The Forest Service can grant grazing rights to ranchers and mineral rights to mining companies. It makes trees available in certain areas to timber companies, which bid for the rights to available tracts.

The national forests were established in 1907, at a time when the forests of the Northwest and elsewhere seemed almost endless. The federal government owned great tracts of land, and the Forest Service was created to manage that land. For many years the forests were managed loosely, and little demand existed for national forest timber. In the late 1940s, after World War II, the country entered a building boom. The demand for wood products was met with increased logging on private and public lands. The redwoods were extremely hard hit. Nearly 85 percent of the redwood forests, most of them in private ownership, fell. Clear-cutting was the common and preferred means of harvesting trees. But the forests suffered. Clear-cuts were ugly, they caused erosion, and they often lay barren. In more recent years, as environmental awareness heightened and new laws were enacted by Congress, the Forest Service became more selective about its management and timber sales. Private forests, too, began to face tougher environmental regulations. Nevertheless, by the early 1990s, as much as 90 percent of the old growth in the national forests of the Pacific Northwest had been cut.

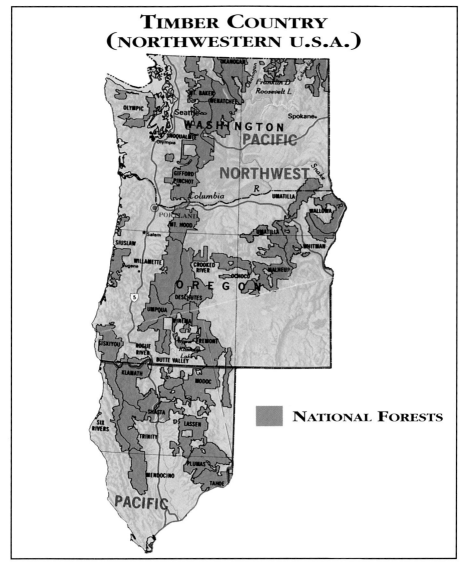

TIMBER COUNTRY
(NORTHWESTERN U.S.A.)

NATIONAL FORESTS

CHAPTER 6

TIMBER AND THE ENVIRONMENT

The cutting of trees for processing into wood products is governed by hundreds of regulations. Public and private forests are subject to environmental laws and safeguards. These are designed to keep the forests healthy and protect the whole forest environment, including air, water and wildlife.

The laws that govern forest management have been sharply questioned in recent years. **Environmentalists** are people with a special concern for the natural environment. Many environmentalists claim that current laws are not strict enough, and that they are not being followed. The timber industry feels that many of the laws are too tough and too confusing. Such laws, says the timber industry, force companies to lay off workers because the companies can no longer cut timber when and where they choose.

The argument about environmental laws and the cutting of timber boiled in the early 1990s in the Pacific Northwest. At the center of the argument was, curiously enough, an owl. Scientists learned that northern spotted owls nested in the old growth forests of Oregon, Washington and California. It was reasonable to assume, the scientists concluded, that the northern spotted owl would become extinct if the ancient forests continued

The northern spotted owl, a member of the old growth forest community in the Northwest, began to disappear along with the ancient forests

to fall. The United States Fish and Wildlife Service, part of the Department of Interior, agreed. It listed the owl as a "threatened" species under the provisions of the Endangered Species Act of 1972.

The Endangered Species Act was passed to protect animals that the Fish and Wildlife Service determined were "threatened" or "endangered." Protection under the laws of the act includes safeguards for an animal's habitat. Suddenly the old growth forests, homes of the northern spotted owl, were closed to logging. The owls paid no attention to the news, of course, but humans reacted loudly.

Many sawmills closed during the early 1980s in timber country, and debate over logging and environmental welfare created more problems for timber interests in the 1990s

The owl's new status – an animal threatened with extinction – divided communities in timber country. Loggers, who had faced hardships during a sluggish timber market in the early 1980s, found their jobs threatened again. Signs appeared that said, "Save a Job, Eat an Owl."

Some environmental groups, however, were delighted with the new restrictions on logging. Their position was that the Pacific Northwest had lost enough of its ancient forests and could not afford to lose the rest.

The issue carried beyond the spotted owl and the **marbled murrelet,** another threatened bird of the mossy old forests. The issue went beyond the three species of Pacific salmon whose spawning streams were sometimes choked by the debris and mud of logging activity. The issue, in fact, went far beyond timber country. It became a national debate about the future of the Endangered Species Act and the future of the remaining ancient forests themselves.

The argument continues. Now it is up to the politicians and judges – not a wise old owl – to settle some of the questions about wildlife, the environment and forest management. It is likely that no one will be completely satisfied with the outcomes.

It is probable that, in the 21st century, companies with tree farms stand the best chance of continued success. Timber companies and mills that rely on being able to cut timber on public lands, such as the national forests, may find themselves with fewer places to harvest trees.

CHAPTER 7

VISITING TIMBER COUNTRY

If timber country were a book, it would read like a catalog of natural wonders, and it would be a catalog as thick as a redwood. The six national parks in timber country – Mount Rainier, North Cascades, Olympic, Crater Lake, Redwood and Lassen Volcanic – have some of

The redwood forest is mirrored in a pool along Redwood Creek in Redwood National Park

the best pages in the catalog. Mount Rainier, an easy drive from Seattle, is the home of 26 active glaciers. With its high, flowering meadows and dense woodland, it is a hiker's paradise, like much of the Pacific Northwest. Olympic National Park, on Washington's Olympic peninsula, protects wilderness beaches, coastal and mountain forests, ice caves and alpine meadows where the whistles of Olympic **marmots** echo. Rain forests shadow the park's Queets, Quinault and Hoh Rivers. Redwood National Park in northwestern California preserves some of the finest remaining stands of old growth redwoods. The oldest redwoods, tucked into foggy coastal valleys with wild gardens of ferns and rhododendron, are 2,000 years old.

Hundreds of state parks and protected sites within the national forests reveal the natural beauty of timber country, too. The scenic, 350-mile Oregon coast is almost entirely in public ownership. The many state parks and national forest lands along the coast encourage public use. One of the highlights is the Cape Perpetua Visitor Center of the Siuslaw National Forest near Yachats, Oregon. Trails from the visitor center snake down to seaside tide pools and climb into evergreen forests above the rugged cape.

Rain falls on timber country throughout the late fall, winter and well into spring. Most visitors don't need water wings, though, because they arrive in summer. Summer days are generally bright, except for bouts of morning and late afternoon fog along the coast. Only residents of timber country know the gloomy spells, but winters, however gloomy, are generally mild and snow-free. And a bit of drizzle is a small price to pay for the greatest forests on earth.

Nisqually Glacier, an ancient, frozen river of ice, sprawls on a slope in Mount Rainier National Park

GLOSSARY

carnivore – a meat-eating animal

choker – a device that holds a log while a cable attached to the choker retrieves the log

clear-cut – having been cleared of trees in a logging operation; the site itself

conifer – a tree that bears seeds in cones, especially the needle-leaved trees

deciduous – the characteristic of a plant periodically losing nearly all of its leaves

environmentalist – someone with a special concern for the natural environment

grapple loader – the device that hoists logs from a platform to a log truck

habitat – the immediate surroundings of an animal or plant; the plant's or animal's specific type of living area

marbled murrelet – a small diving seabird that nests in the Pacific Northwest

marmot – a large ground squirrel of the Western mountains; the Western relative of the woodchuck or groundhog

niche – an organism's place, role or job in the community

nutrients – substances providing an organism with nourishment

revenue – income from a job, investment or other source

scaler – one who measures and rates logs for value

GLOSSARY

second-growth – a forest that grows in place of the one which has been destroyed

seedlings – small, young trees

silviculture – the science of harvesting and planting trees

skidder – a bulldozerlike vehicle that uses steel jaws to haul logs from a cutting site

temperate – referring to that part of the earth in the Northern Hemisphere between the Tropic of Cancer at 23 1/2° north of the equator, and the Arctic Circle at 66° north of the equator

timberline – the place above which no trees can grow because of cold, wind, and poor quality or frozen soil

INDEX

Alaska 13
animals 11, 21, 36, 39
bear, black 11
Bend, OR 6
birds 19
British Columbia 11, 13, 16
cable yarder 26
California 5, 6, 13, 17, 33, 34, 36, 38, 43
Cape Perpetua Visitor Center 43
carnivores 21
Cascades 6
cedar, western red 13
chainsaw 25
choker 26
clear-cutting 36
clear-cuts 36
coastal forest 13, 14, 16, 17
Congress 37
Coos Bay, OR 23, 24
cougars 11
Crater Lake National Park 42
dairy farms 6
Department of Agriculture (U.S.) 34
Department of Interior (U.S.) 39
disease 19
docks 23, 26
Douglas-fir 6, 13, 16, 26, 33
elk, Roosevelt 11
Endangered Species Act 39, 41
environment 38, 41
environmentalists 38
evergreens (see *conifers*)
faller 25
ferns 19
Fish and Wildlife Service (U.S.) 39
fog 43
food chain 22
food energy 21, 22
food web 22
forest, ancient (see *forests, old growth*)

Forest Service (U.S.) 34, 36, 37
foresters 32
forests 5, 6, 8, 11, 13, 19, 25, 26, 31, 32, 33, 36, 37, 38, 43
old growth 19, 21, 37, 38, 39, 40, 41
second-growth 19
Georgia 11
glaciers 6, 43
grapple 26
grapple loader 26
habitats 11, 39
helicopters 26, 33
hemlock, western 6, 13, 33
Hoh River 43
insects 19
jobs 9, 11
Lassen Volcanic National Park 42
lightning 19
loggers 40
logging 25, 36, 39
logging trucks 23, 25
logs 25, 26
lumberjack 25
marmots 43
mills (see *sawmills*)
moss 16
Mount Rainier National Park 42, 43
mountains 6, 14
murrelet, marbled 41
national forests 33, 36, 41, 43
Newport, OR 23
North Cascades National Park 42
ocean currents 14
Olympia, WA 6
Olympic National Park 42, 43
Olympic peninsula 16, 43
orchards 6
Oregon 5, 6, 9, 11, 13, 15, 34, 38, 43
owl, spotted 38, 39, 40, 41
owls 11, 22

INDEX

Pacific Northwest   5, 6, 25, 37, 38, 40, 43
Pacific Ocean   6, 13
paper   9
pines   13
pitcher plant, California   36
plants   11, 21, 22, 36
Portland, OR   6
precipitation   14, 15, 16, 43
Queets River   43
Quinault River   43
rain (see *precipitation*)
rain forest, temperate   16, 43
ranchers   36
Redwood National Park   42, 43
redwoods   6, 17, 19, 33, 36, 43
reforestation   32
Salem, OR   6
salmon   41
sawmills   26, 41
scaler   26
seashores   6
Seattle, WA   6, 43
seedlings   32
seeds   21, 32
shrubs   19
Sierras   6
silviculture   33
silviculturists   33
Siuslaw National Forest   43
skidder   26
slug, banana   11
spruce, Sitka   6, 13, 16
state parks   43
squirrel, Douglas   22
sunlight   21

Tacoma, WA   6
timber   11, 15, 31, 32, 36, 38
timber industry   38
tree farmers   33
tree farms   33, 34, 41
trees   11, 13, 19, 21, 23, 25, 31, 33, 36, 38, 41
  conifers   13, 17, 22
  deciduous   13
Waldport, OR   6
Washington   5, 6, 11, 13, 16, 34, 38, 43
waterfalls   6
wilderness areas   36
wildflowers   19
wood   9, 11, 29
wood products   9, 29, 36, 38
wood pulp   9
Yachats, OR   43